4th Grade Science Volume 5

© 2013 Todd Deluca
OnBoard Academics, Inc
Newburyport, MA 01950

800-596-3175
www.onboardacademics.com

Table of Contents

Light and Color   3

Light and Color Quiz   12

Energy and Work   13

Energy and Work Quiz   23

Transferring Energy   24

Transferring Energy Quiz   33

Electric Circuits   34

Electrical Circuits Quiz   44

Light and Color

How does a rainbow form?

You've probably seen a rainbow form when the sun comes out on a rainy day. But why does the combination of the sun and raindrops produce a rainbow?

What is your idea about why the sun and raindrops form a rainbow?

_____

_____

_____

Making a Rainbow with a Flashlight and a Prism

White light is actually made up of several different colors, and we can use a transparent prism to break up white light to observe the spectrum of colors within it. This is similar to observing a rainbow. Raindrops in the air act like tiny prisms that break up the light passing through them, creating the colorful rainbows that we see on the horizon.

Now that you've learned about prisms can you explain why sun and raindrops form a rainbow?

_____

_____

_____

What is the visible light spectrum?

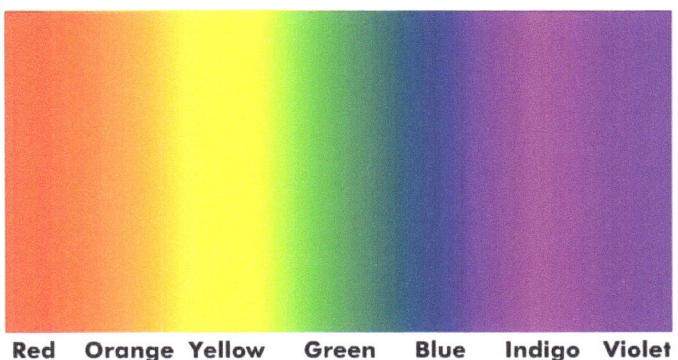

Red Orange Yellow Green Blue Indigo Violet

When broken apart, white light separates into the same spectrum of colors in order of wave length. The order of colors by wave length is; red, orange, yellow, green, blue, indigo and violet.

A good way to remember this is to think of an acronym such as Roy G. Biv.

The wave length of a color tells you how much energy it has. For example red light has the least amount of energy and has the longest wave length. Violet has the most amount of energy and so it has the shortest wave length.

> **White light contains a spectrum of colors that appear in order of their wavelengths. From longest wavelength to shortest wavelength (or least energy to most energy), they are red, orange, yellow, green, blue, indigo and violet (or ROY G. BIV).**

Order the spectrum of colors by wave length.

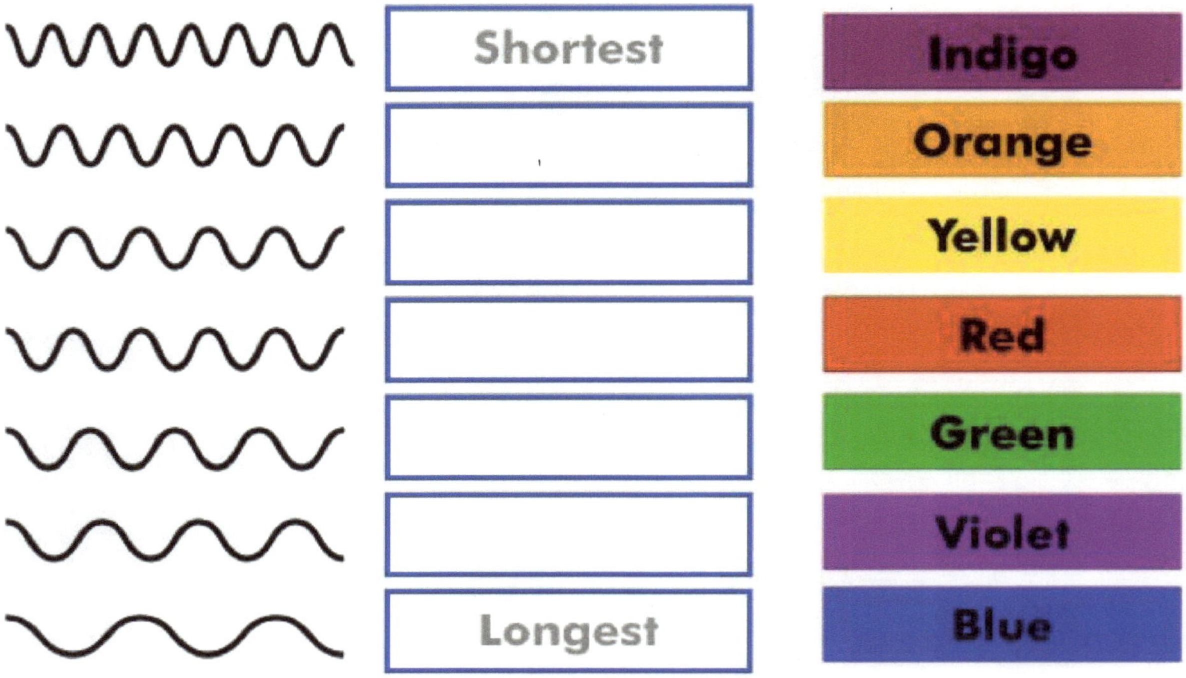

∿∿∿∿∿∿∿	**Shortest**
∿∿∿∿∿∿	
∿∿∿∿∿	
∿∿∿∿∿	
∿∿∿∿	
∿∿∿∿	
∿∿∿	**Longest**

Indigo

Orange

Yellow

Red

Green

Violet

Blue

How is light related to color?

Objects appear to be a certain color because of the colors they absorb and the colors they reflect. For example look at this red rose.

When light is present, all colors of the spectrum are shining on the red rose; red, orange, yellow, green, blue, indigo and violet. The rose absorbs orange, yellow, green, blue, indigo and violet. However the red light is not absorbed and bounces off into our eye and that is why we see the rose as red.

When no light is present, there is no reflection of light to reflect off of objects into our eyes so nothing is visible.

Which colors are being observed and which colors are being reflected for each item.

X for absorbed

√ for reflected

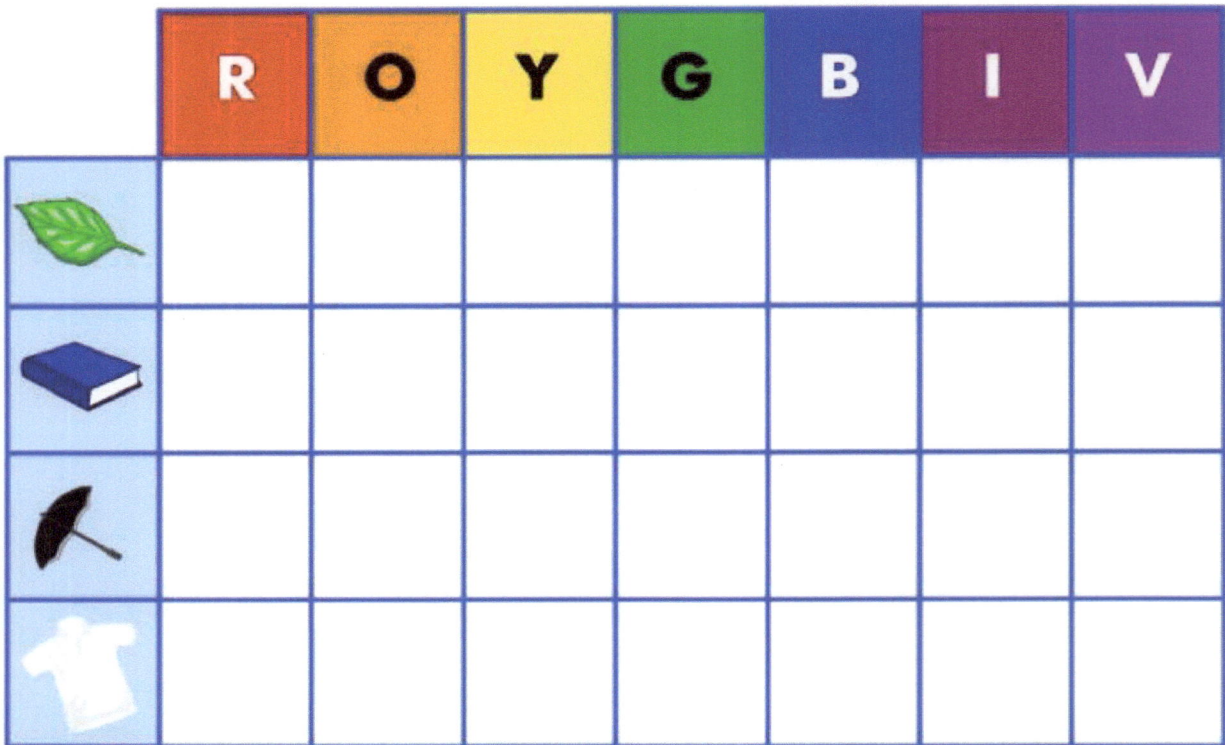

	R	O	Y	G	B	I	V
🍃							
📘							
☂							
👕							

Primary Colors

Almost any color can be created by mixing the three primary colors of light; red, green and blue.

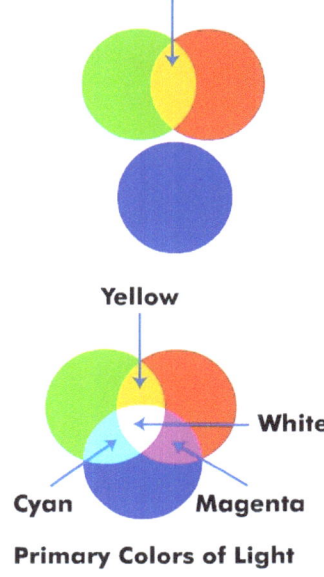

For example when green and red overlap we see yellow. When all three colors overlap we see white light. This only works when mixing light.

When mixing paint, the primary colors are yellow, cyan and magenta. When all three paint colors are mixed we get black.

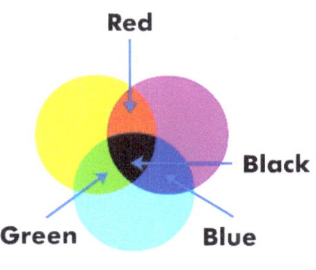

    www.onboardacademics.com

What colors are created when you mix the following color combinations.

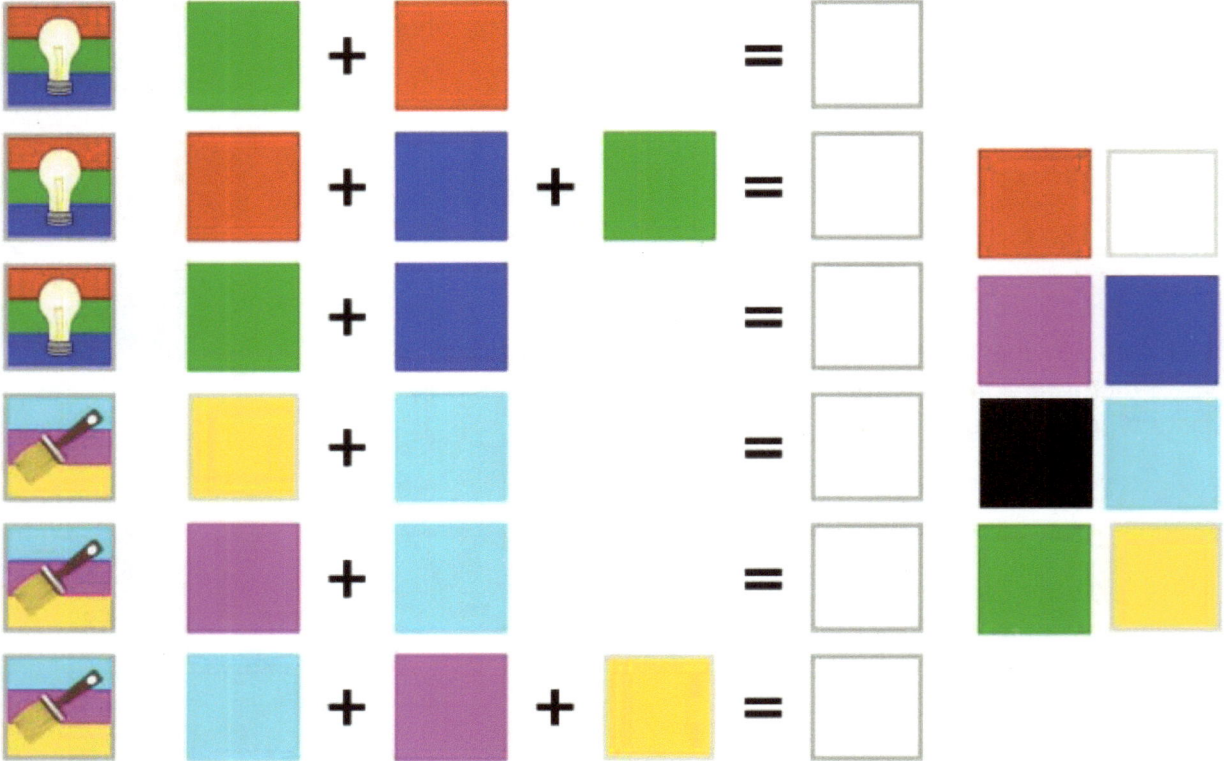

Light and Color Quiz

1. For a rainbow to be seen, you need light from the sun and raindrops in the air. True or false.

2. White light is made up of _____ colors
    - a. 8
    - b. 7
    - c. 6
    - d. 10

3. Raindrops in the air act like tiny prisms that break up light passing through them, creating rainbows. True or false?

4. When broken apart, white light splits into a spectrum of colors in order of _____.
    - a. brightness
    - b. wavelength

5. What is the acronym to remind you of the order of colors?
    _____

6. What color has the least energy? _____

7. What color has the most energy? _____

8. What color do you get when you mix the three primary colors of light? _____

Energy and Work

What are energy and work?

Energy causes things to happen all around us. In science we say that energy has the ability to cause motion or to create change. For example, the energy we get from food helps us to grow and to do things like to kick a soccer ball.

When we put gasoline in the car we are using the gasoline to create chemical energy to create motion energy to turn the wheels of the car.

When we are at home and turn on a light, we are converting electrical energy into light energy.

When we kick a ball, drive a car or turn on a light switch, we say that energy is doing work. In science, work is defined as the transfer of energy that occurs when an object is moved a distance, or when something undergoes a chemical change.

So when the boy pushes this table and the table moves in the direction in which the boy is pushing the table we would say that work has been done. If the boy pushes against this giant rock and the rock doesn't move, we would say that no work has been done even though the boy has used a lot of energy and feels as though he has done a lot of work.

You can find energy and work in all sorts of places. The water in this dam has a type of energy that we call potential energy. Potential energy is stored energy. When the dam opens the stored potential energy is transferred into motion energy. Because the water has moved, we say that work has been done.

There are many different forms of energy including chemical energy, motion energy, heat energy and electrical energy.

Most of the energy on Earth comes from the sun. Energy from the sun warms the Earth, gives us daylight and is used in the process of photosynthesis. This is the process of plants making their own food. Sunlight is converted into chemical energy within the leaves of a plant.

Connect the correct energy source to each scene to enable work to be done.
Describe what work is being done in each scene.

 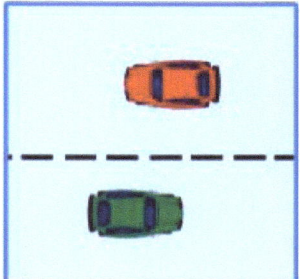 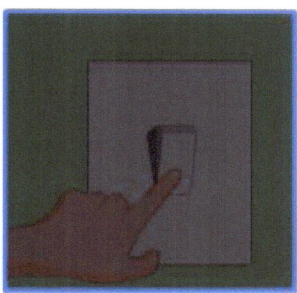 

What work is being done?	What work is being done?	What work is being done?	What work is being done?

Label the different forms of energy.

motion   wind   heat

light   electrical   chemical

When work is being done, energy often changes from one form into another.

When we switch on a light we are using electric energy. The electrical energy is transferred into light energy.

Let's take a look at how electricity is produced and reaches our homes. Along the way, we'll see how energy is transferred into different forms as work is being done.

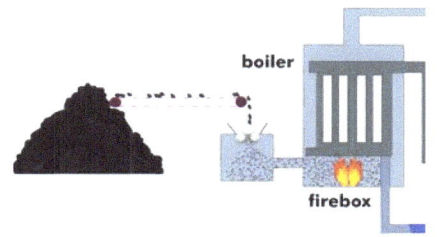

Electric energy usually starts with coal as coal is used in half the power plants within the USA.

When coal is burned in a power plant it heats water in a boiler to very high temperatures creating a lot of steam. At this point work is done because water is heated and turned into steam and energy is transferred because chemical energy from the coal is transferred into heat energy.

The steam is piped from the boiler at very high pressure and used to turn large blades called turbines. At this point work is being done because the turbines turn and energy is transferred because the heat energy from the steam is transferred to motion energy.

The turbine is connected to a generator. The generator has large magnets that are surrounded by coils of copper

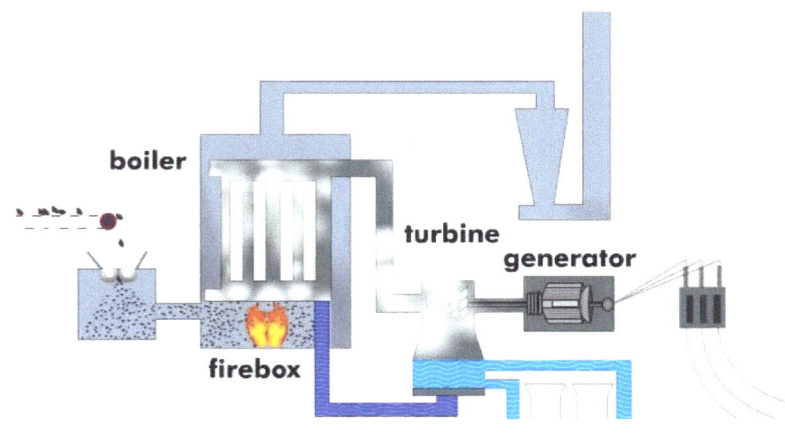

wires. The turbine turns the magnets causing them to spin within the copper coils and electricity is created. Work is done as magnets spin and energy is transferred because the motion energy from the turbine is transferred to electrical energy.

Electricity reaches our home on transmission wires. These are the wires you see on the

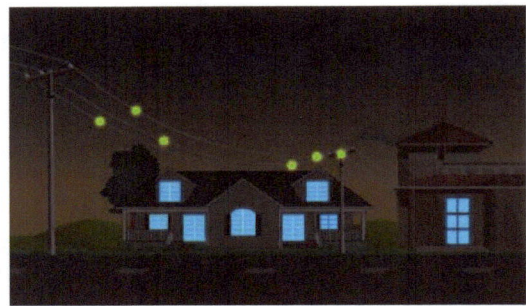

tall poles on the street. Sometimes these wires run underground and are hidden from your view.

Once the energy reaches our home we might transfer it into may different types of energy including heat to cook our food, sound energy to listen to our music or motion energy to run on a treadmill. Whatever the case work is being done!

List 5 types of work done by energy sources in your home.

1. _____

2. _____

3. _____

4. _____

5. _____

Which forms of energy are being transferred in the illustrations below?

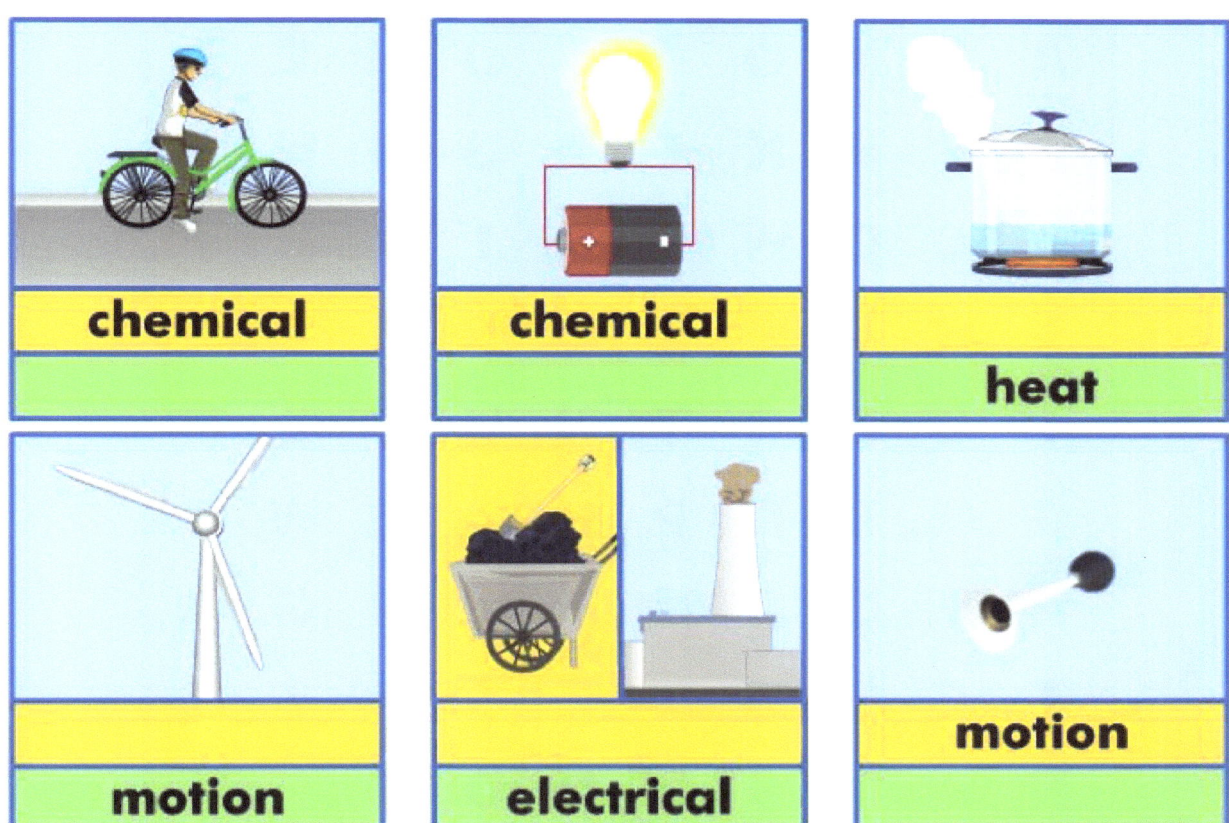

chemical	**chemical**	**heat**
motion	**electrical**	**motion**

sound   electrical   electrical
      wind   motion   chemical

Can you find and circle eight different types of energy?

```
A X M O T I O N C D
Z F B K F K G U N T
A C O W T S I C R F
L H S I I O S L E E
I E B G T N U E E H
G M B D P N D A W E
H I O L O D Z R G A
T C E G T R S R I T
N A T S U N D O R J
E L E C T R I C A L
```

Energy and Work Terms Review

Match!

potential	I'm the energy found in food, gasoline and batteries.
the Sun	I'm the energy that causes music and noise.
sound	I'm stored energy not yet in motion.
chemical	I'm the energy generated at a power station.
work	I'm the source of most energy on Earth.
electrical	This is done when an object moves a distance or a substance undergoes a chemical change.

Energy and Work Quiz

1. Wind is a form of energy. True or false?

2. The faster something moves, the _____ energy it has.
   a. more
   b. less

3. Oil, gas and coal all contain stored _____ energy.
   a. electric
   b. light
   c. heat
   d. chemical

4. A stove uses _____ energy to turn water into steam.
   a. wind
   b. motion
   c. heat
   d. chemical

5. Work is done when a mass is moved a distance. True or false?

6. The energy found in our food is in the form of chemical energy. True or false?

Transferring Energy

Potential and Kinetic Energy

There are six main forms of energy and many different types within each form but all energy exists in one of two states; potential energy and kinetic energy.

 The energy stored in the water that is being blocked by the dam is potential energy. When the dam opens and the water is flowing downstream this energy become kinetic energy.

Potential energy is stored energy while kinetic energy is energy in motion.

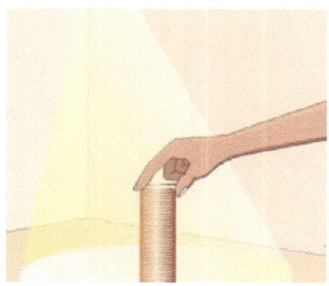 There are many different types of potential and kinetic energy. For example, a coiled spring has potential energy but when the spring is released it become kinetic energy.

A giant rock on top of a hill has potential energy but when it rolls down the hill the potential energy is transferred to kinetic energy.

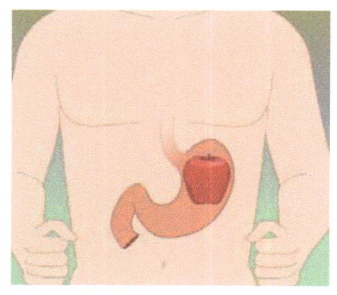 The food that is stored in our body as chemical energy is potential energy but it is converted to kinetic energy every time that we move.

There are six main *forms* of energy, (and many different types of energy within each form), but energy exists in two states: potential energy and kinetic energy. Potential energy is stored energy, while kinetic energy is energy in motion.

     www.onboardacademics.com

Potential energy can change into kinetic energy and vice versa.

Examine the rolling ball.

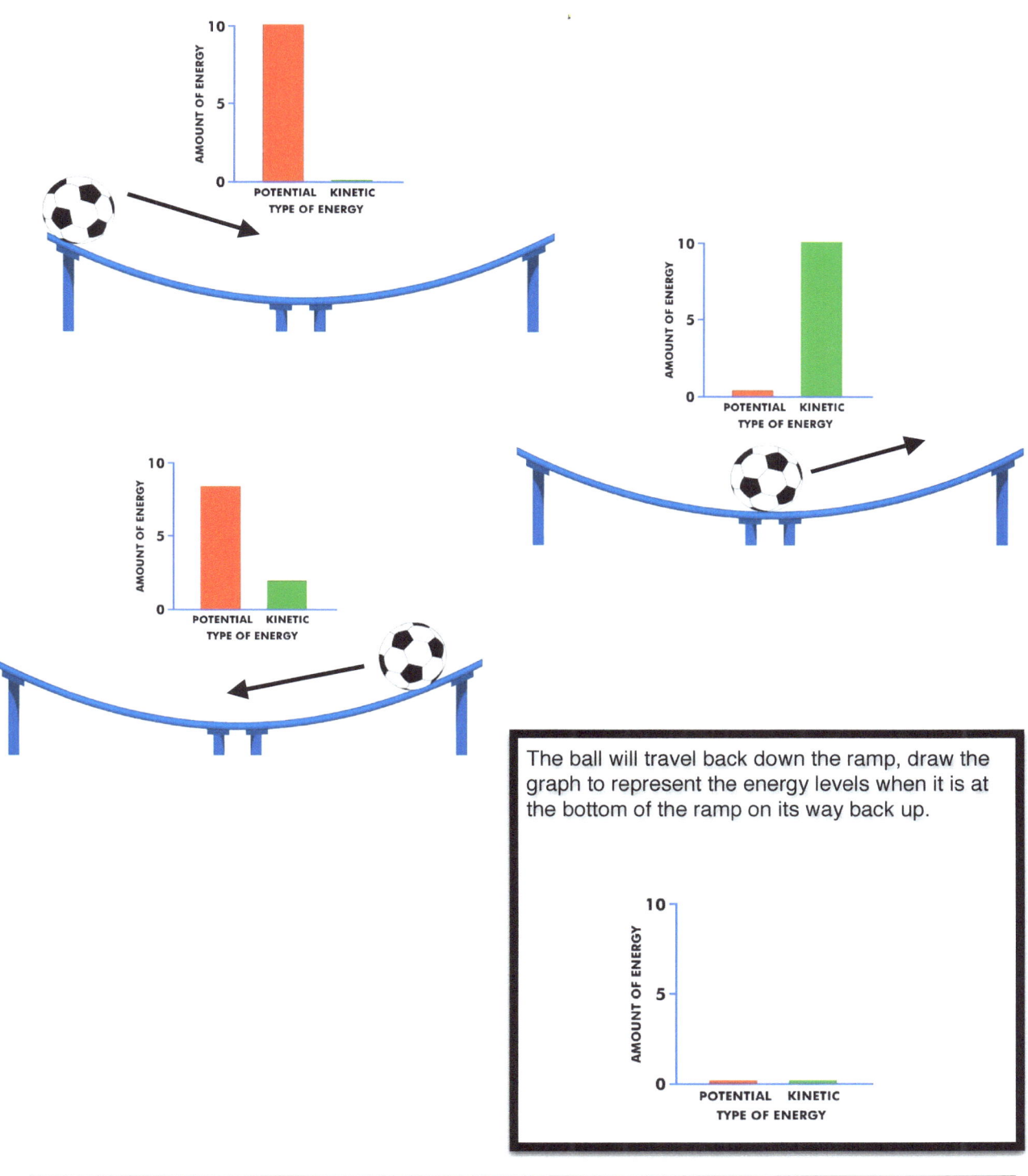

The ball will travel back down the ramp, draw the graph to represent the energy levels when it is at the bottom of the ramp on its way back up.

Identify the energy status on the swing set.

1 Kinetic energy is being transferred into potential energy

2 Potential energy is being transferred into kinetic energy

3 Potential energy is zero; kinetic energy is at its maximum

4 Kinetic energy is zero; potential energy is at its maximum

Friction and Heat Energy

Will this ball ever stop rolling

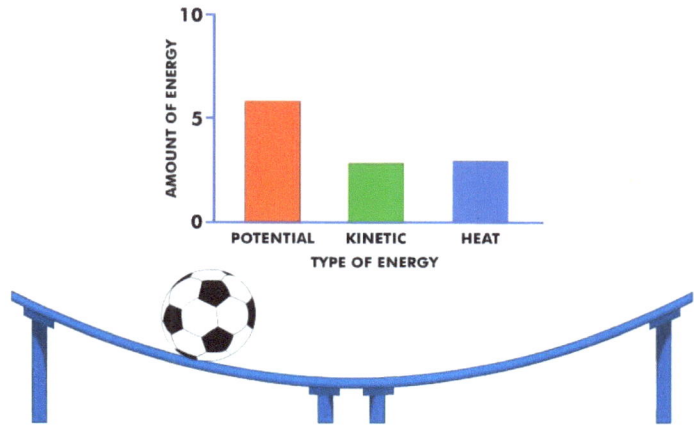

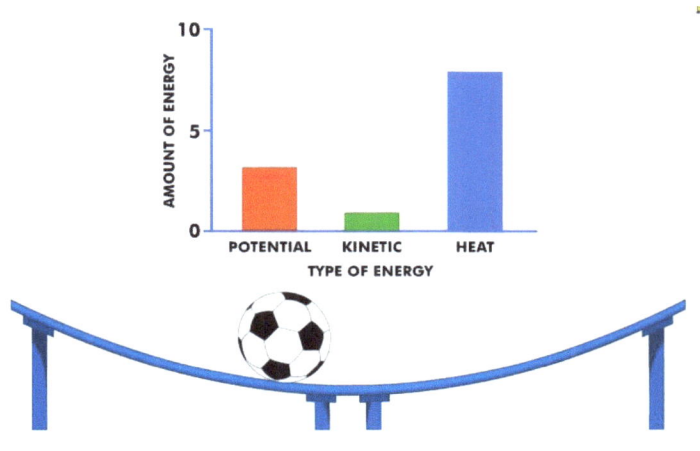

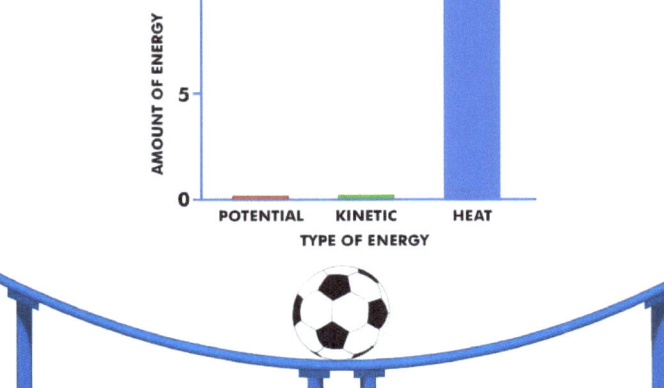

The soccer ball will eventually stop moving because there is **friction** between the ball and the ramp. When the ball rubs against the ramp, some of the ball's kinetic energy is transferred into **heat energy** in the ramp. The ball slows down, and the ramp heats up a little bit.

Identify the 6 forms of energy.

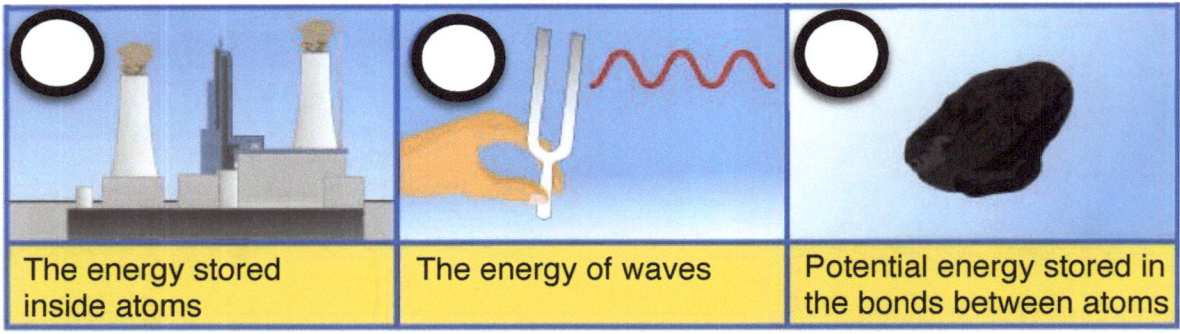

| The energy stored inside atoms | The energy of waves | Potential energy stored in the bonds between atoms |

(C) **chemical**   (T) **thermal**   (W) **wave**

(M) **mechanical**   (N) **nuclear**   (E) **electrical**

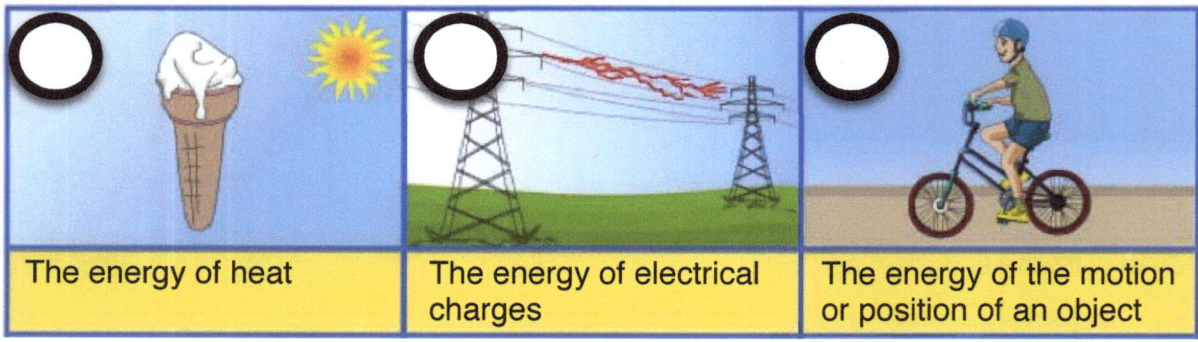

| The energy of heat | The energy of electrical charges | The energy of the motion or position of an object |

www.onboardacademics.com

Organize this energy web.

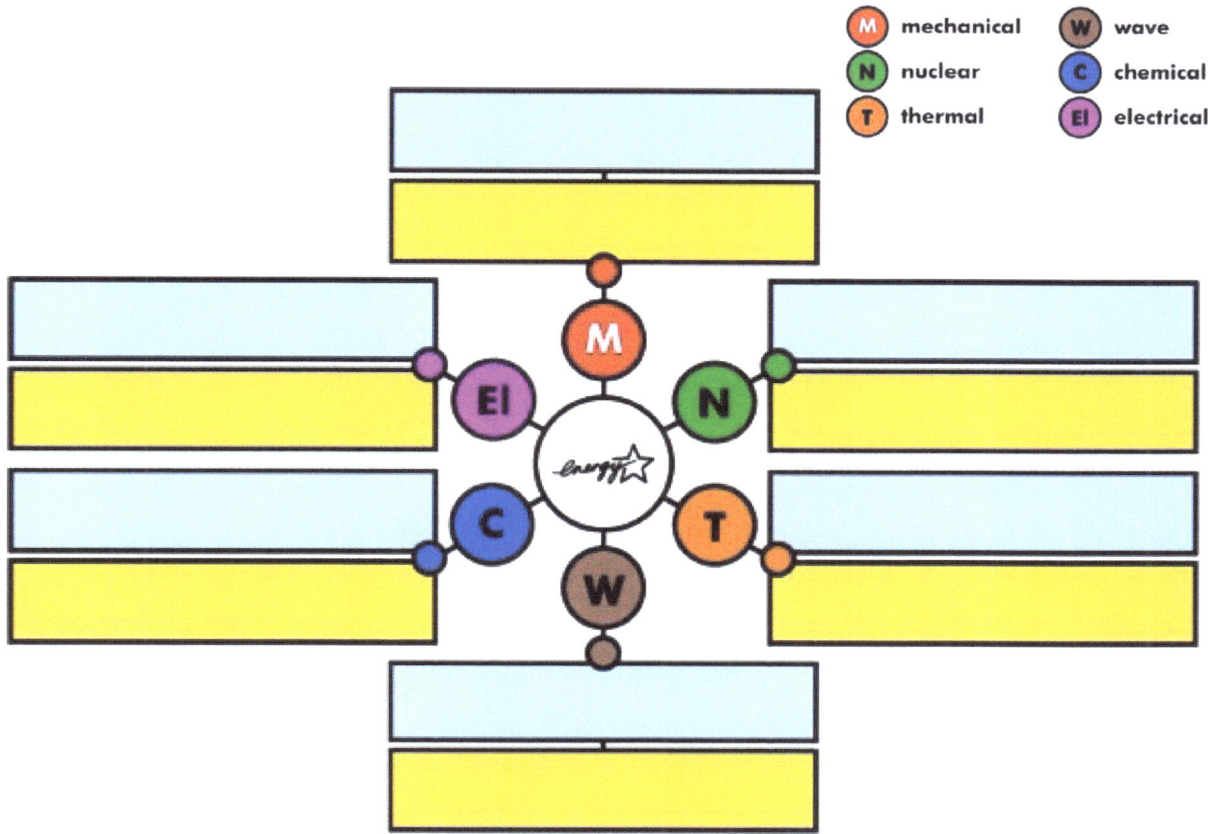

Directions

Write the description for the type of energy in the empty blue boxes in the energy web using the suggestions in blue boxes.

Write the examples of the type of energy in the empty yellow boxes by using the suggestions provided in the yellow boxes.

a moving bicycle, jumping frog, a hanging weight

static shocks, lightning power lines

food, matches

ice cream melting

nuclear power plants, the Sun's core

microwaves, visible light, ultraviolet radiation

energy inside atoms

energy of light and radiation

energy of the motion or position of an object

energy of electric charges

energy of heat

potential energy stored in bonds between atoms

Energy transfer is what makes all work possible.

Batteries don't make energy, they convert chemical energy into electrical energy. The metals inside batteries react with an acid or with a base that release electrons that can flow through a wire and make something work.

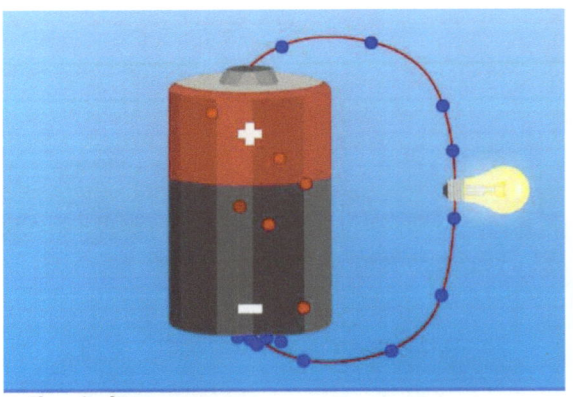

● Chemical energy
● Electrical energy

Photosynthesis is an example of energy transfer. The electro magnetic energy of sunlight is a form of wave energy that is transferred into chemical energy during the process of photosynthesis.

When you sharpen a pencil, you body transfers the chemical energy, that is derived from the food you eat, into mechanical energy. The energy is transferred to your arm and the pencil sharpener.

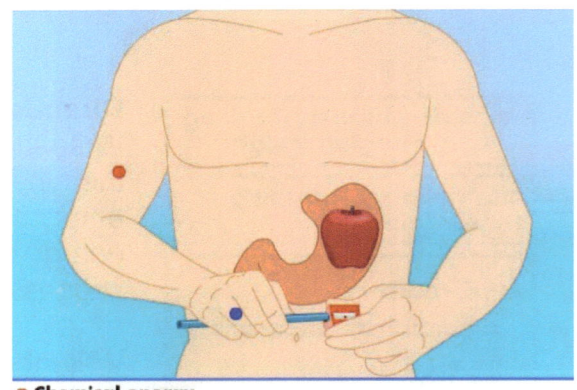

● Chemical energy
● Mechanical energy

Identify the type of energy transfer.

A stove top transfers

_____ **energy into**

_____ **energy.**

A cyclist transfers

_____ **energy into**

_____ **energy.**

A light bulb transfers

_____ **energy into**

_____ **energy.**

electrical   chemical   wave

mechanical   electrical   thermal

Transferring Energy Quiz

1. Potential energy is stored energy. True or false?

2. A moving train is a good example of which of the following forms of energy.
    a. potential energy
    b. kinetic energy
    c. electrical Energy

3. _____ is the potential energy stored in the bonds between atoms.
    a. chemical energy
    b. nuclear energy
    c. thermal energy

4. Electromagnetic energy is the energy of electrical charges. True or false?

5. The energy stored inside atoms is _____.

6. The thermal energy of sunlight is converted into mechanical energy during the process of photosynthesis. True or false?

Electric Circuits

The parts of a simple electric circuit.

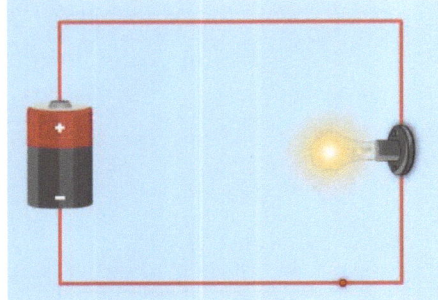

The elements that are required to make a bulb glow. They are the bulb itself and the appliance, a battery which is the power source and a wire, the path to conduct the electricity.

When all of these three elements are in place we call this a circuit. A circuit enables the energy to flow from the minus or negative end of the battery to the plus or positive end of the battery.

If we want to turn the light bulb on and off we must add a switch to the circuit to interrupt the flow of electricity. When the switch is in the on position the circuit is complete and the energy can flow free throughout the circuit making the bulb glow. We say that the circuit is closed.

When the switch is off the circuit is broken and electricity can not flow so the bulb does not glow. In this case, we say that the circuit is open.

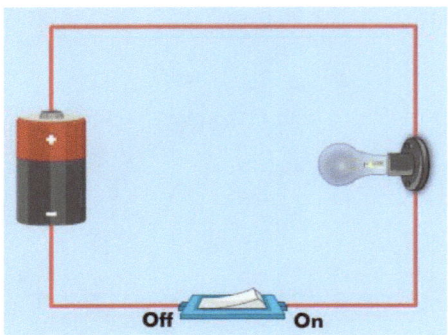

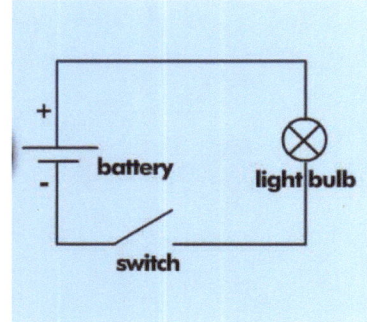

Electric circuits are represented using a circuit diagram. The battery is represented by a big line and a small line. The lightbulb is represented by a circle with an X inside of it. The switch is indicated by a line indicating an open or closed position.

Although electrons (tiny particles that carry an electric current) travel from the negative (-) terminal to the positive (+) terminal, the standard way to show the flow of an electric current is from positive to negative.

Which of these circuit is complete?

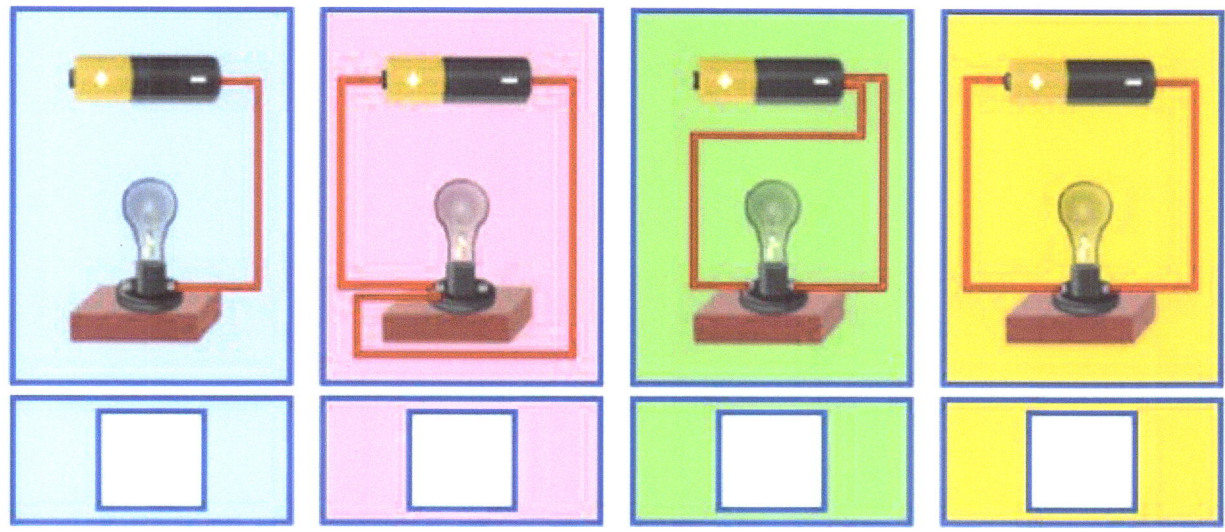

In order for electricity to pass through a circuit, the circuit must be complete. This means that electricity must be able to flow from the negative end of the battery, through the bulb and then back to the positive end of the battery.

Identify the parts of a circuit.

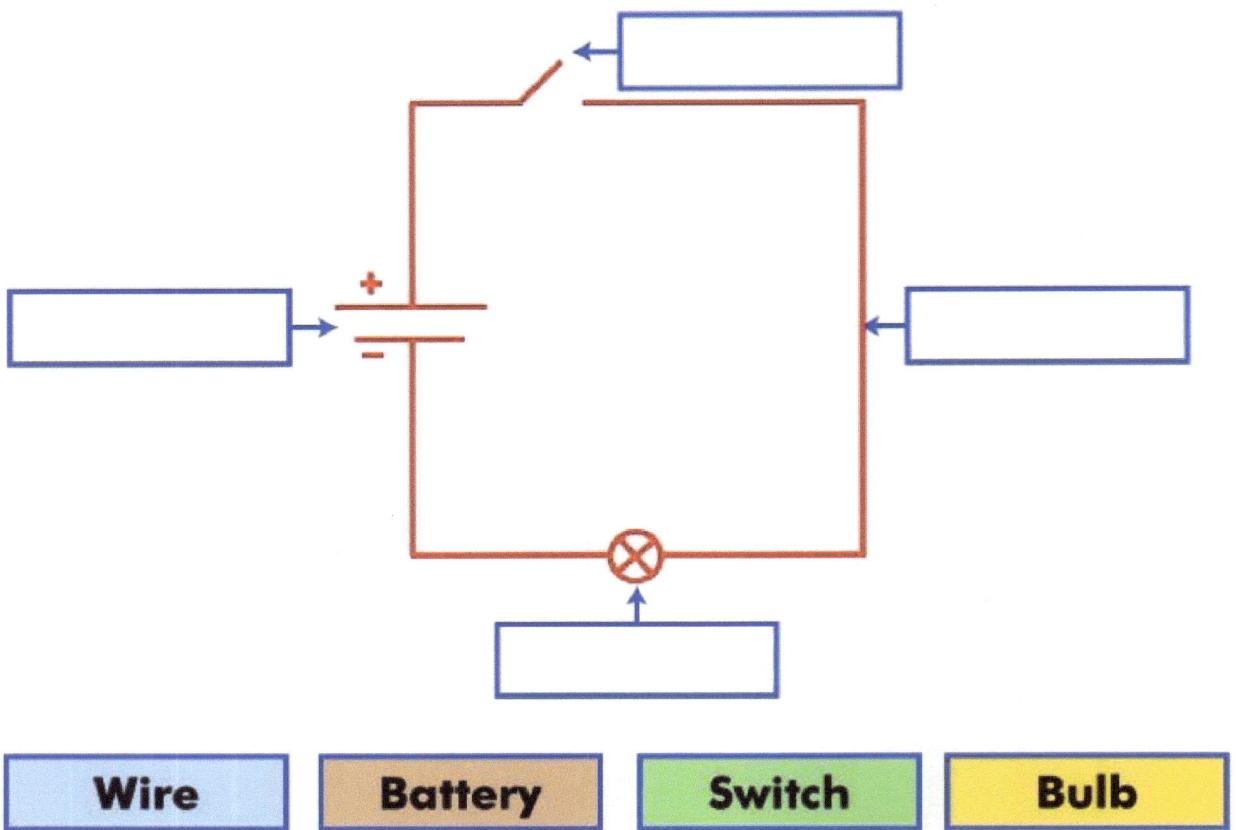

| Wire | Battery | Switch | Bulb |

The Effect of Changes to a Circuit

When you make changes to a circuit this can affect other elements of a circuit. For example if we were to add more wire, less electricity will reach the lightbulb causing a dimmer glow.

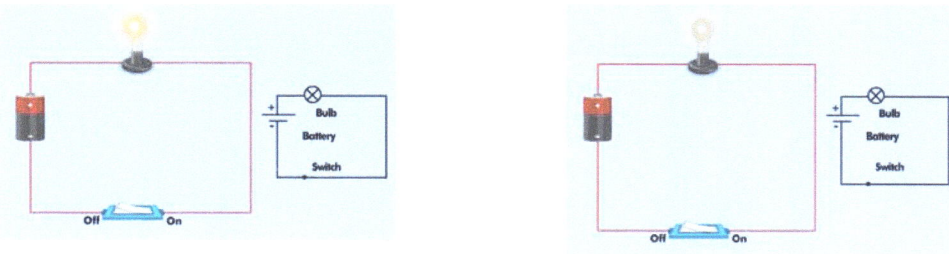

If we were to add another light bulb this too would reduce the intensity of the glow.

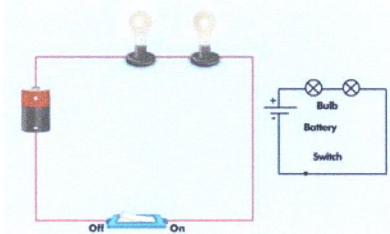

On the other hand, if we were to add another battery, this will increase the intensity of the lightbulbs' glow.

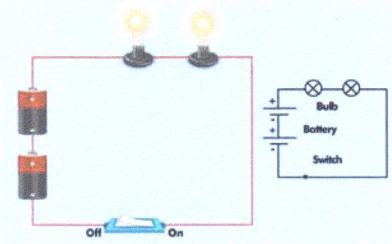

The effective changes to a circuit often depend on what type of circuit it is. Lets explore two types of circuits; series circuits and parallel circuits.

Series Circuits.

There are two main types of circuits; series and parallel.

A series circuit is one in which light bulbs are added in a row so that electricity has to flow through one light bulb in order to get to another.

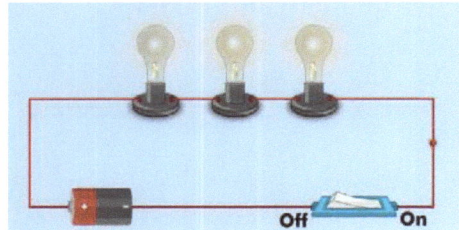

Since the electricity flows through one shared pathway the intensity of the glow decreases as more light bulbs are added to the circuit as electricity meets more resistance.

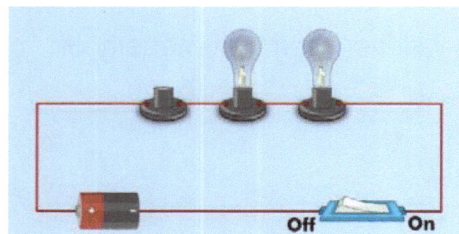

If a lightbulb in a series circuit breaks or is removed, all the lightbulbs will stop working because the circuit is broken and the electricity will not be able to flow in a complete loop.

In a series circuit, bulbs are added in a single row so that electricity has to flow through one bulb in order to get to the next bulb. Since the electricity flows along one shared pathway, the intensity of the glow decreases as more light bulbs are added, and the circuit is broken if a bulb breaks or is removed.

Build a series circuit.

Draw the elements of the circuit onto the wire to represent a series circuit.

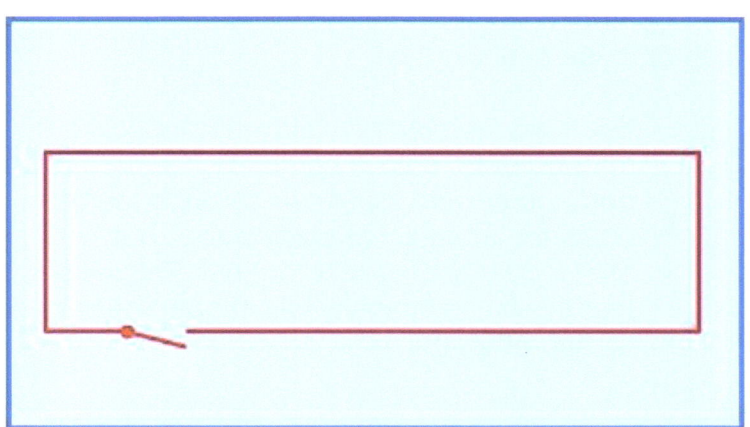

Battery

Bulb

Switch

The intensity of the glow decreases when more bulbs are added to the circuit. As you add batteries, the intensity of the glow increases equally for all the bulbs. However, too much electricity (such as two batteries for one bulb) could make a bulb burn out.

Parallel Circuits

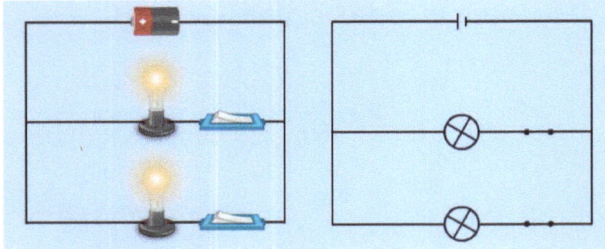

A parallel circuit has the same parts as a series circuit but it's constructed a little bit differently.

In a parallel circuit the electricity still flows from the negative to the positive end of the battery but the bulbs are added in separate parallel rows rather than in a single row.

This means that each bulb has its own loop or circuit within the overall circuit and electricity flows separately through each of these mini circuits.

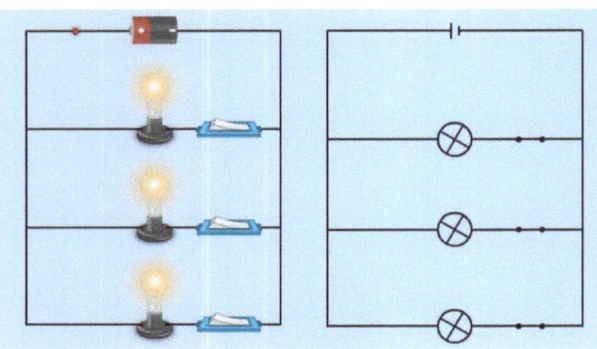

Because the electrical current that flows through each bulb is completely separate from the current that flows through the other bulbs the intensity of each bulb is not affected by adding other bulbs. This means that the bulbs intensity will not decrease by adding other bulbs.

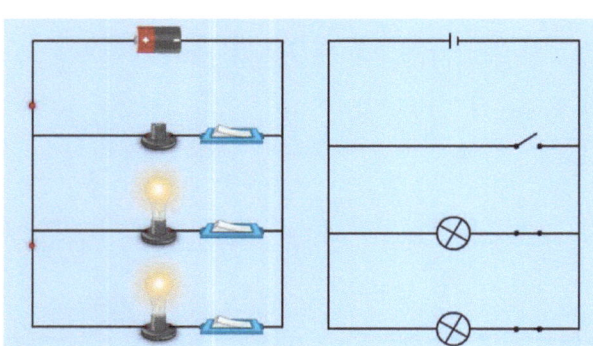

Also, if one bulb were to break or is removed, the other bulbs would continue to work since the other bulb's circuits remain closed.

It's for these reasons that parallel circuits are widely used.

> **In a parallel circuit, bulbs are added in separate parallel rows rather than in a single row. Because each bulb has its own loop, or circuit, the intensity of the glow of a bulb is not affected by adding more bulbs, and if a bulb breaks or is removed, the other bulbs will continue to work.**

Build a parallel circuit.

Draw the elements of a parallel circuit onto the wire diagram.

Battery

Bulb

The intensity of the glow is equally distributed across all the bulbs in a parallel circuit and as you add batteries, the intensity of the glow increases for all the bulbs equally. However, too much electricity per bulb (such as two batteries for one bulb) could make a bulb burn out.

Series or Parallel Circuit?
Label each circuit fact with a P or an S.

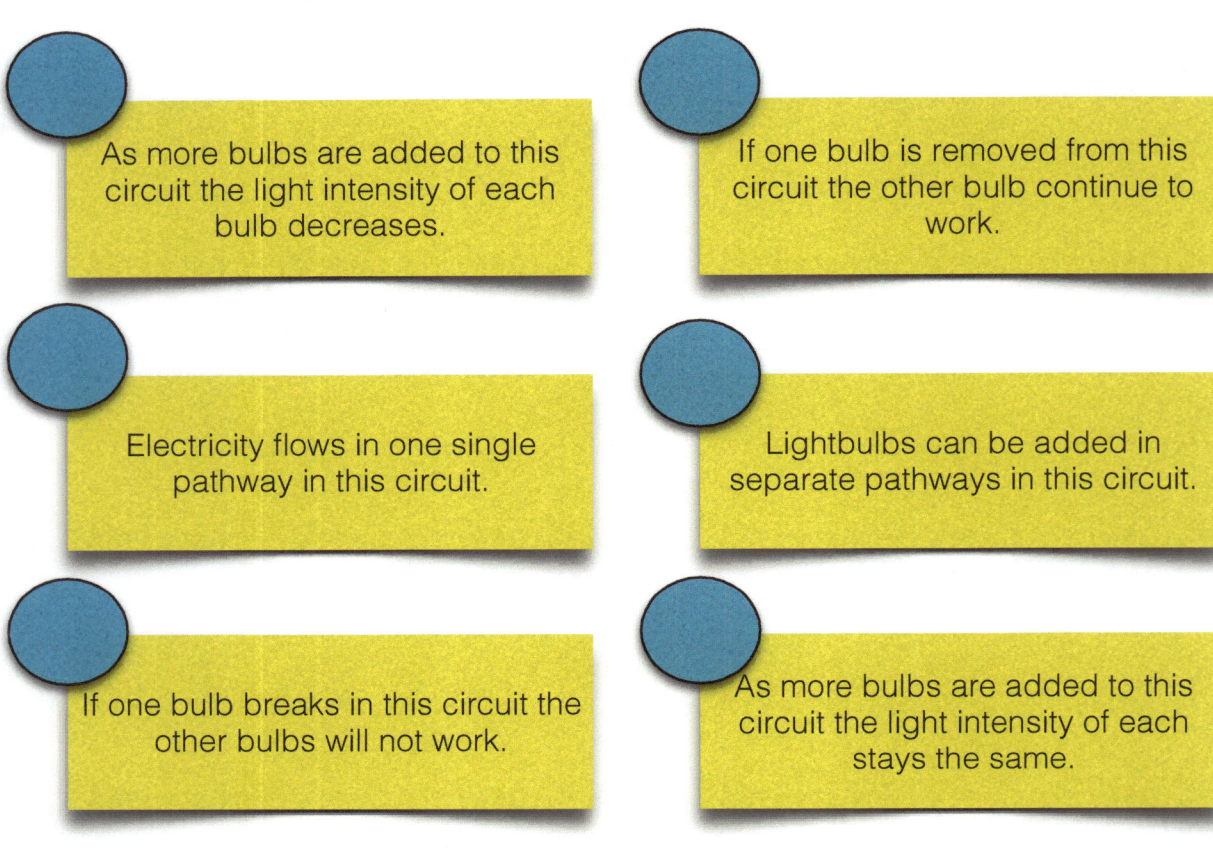

As more bulbs are added to this circuit the light intensity of each bulb decreases.

If one bulb is removed from this circuit the other bulb continue to work.

Electricity flows in one single pathway in this circuit.

Lightbulbs can be added in separate pathways in this circuit.

If one bulb breaks in this circuit the other bulbs will not work.

As more bulbs are added to this circuit the light intensity of each stays the same.

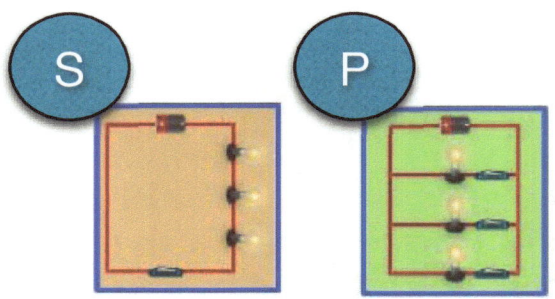

S   P

Electrical Circuits Quiz

1. Electric current is the flow of _____.
   a. protons
   b. neutrons
   c. electrons
   d. atoms

2. A _____ is a device used to open and close a circuit.
   a. bulb
   b. cell
   c. switch
   d. wire

3. As long as a batter and wire are connected in a circuit, current will flow even if there is no switch. True or false?

4. In a circuit, the flow of electrons is from the positive terminal to the negative terminal. True or false?

5. In a series circuit, electricity travels through _____.
   a. one path
   b. many paths
   c. two different paths

Newburyport, MA 01950

1-800-596-3175

OnBoard Academics employs teachers to make lessons for teachers! We create and publish a wide range of aligned lessons in math, science and ELA for use on most EdTech devices including whiteboard, tablets, computers and pdfs for printing.

All of our lessons are aligned to the common core, the Next Generation Science Standards and all state standards.

If you like our products please visit our website for information on individual lessons, teachers licenses, building licenses, district licenses and subscriptions.

Thank you for using OnBoard Academic products.
